Amanda Rosa Olivier Hurtado

Integrated agroecological production systems Savanna ecosystem

AF667114

Amanda Rosa Olivier Hurtado

Integrated agroecological production systems Savanna ecosystem

Experience of the Endogenous Development Nucleus "19 de Abril"

ScienciaScripts

Imprint
Any brand names and product names mentioned in this book are subject to trademark, brand or patent protection and are trademarks or registered trademarks of their respective holders. The use of brand names, product names, common names, trade names, product descriptions etc. even without a particular marking in this work is in no way to be construed to mean that such names may be regarded as unrestricted in respect of trademark and brand protection legislation and could thus be used by anyone.

Cover image: www.ingimage.com

This book is a translation from the original published under ISBN 978-620-0-01713-0.

Publisher:
Sciencia Scripts
is a trademark of
Dodo Books Indian Ocean Ltd. and OmniScriptum S.R.L publishing group

120 High Road, East Finchley, London, N2 9ED, United Kingdom
Str. Armeneasca 28/1, office 1, Chisinau MD-2012, Republic of Moldova, Europe
Printed at: see last page
ISBN: 978-620-7-40551-0

Copyright © Amanda Rosa Olivier Hurtado
Copyright © 2024 Dodo Books Indian Ocean Ltd. and OmniScriptum S.R.L publishing group

THANK YOU

To my Almighty God for giving me the wisdom and patience to achieve this accomplishment.

To my beautiful twins Jesus Daniel and Jose Daniel for their patience and support at all times during the development of my studies.

To Robert Heredia for his technical guidance.

To my partners Jorgelina and Elizabeth for their good contributions.

To the members of the Kamaropa cooperative for their valuable collaboration and help.

To the engineer Arnaldo Guevara for his unconditional help and his human qualities.

A thousand thanks to all of you.

Design of a proposal for an integrated agricultural production system as an agroecological alternative in savanna ecosystems
as an agroecological alternative in savanna ecosystems, in the Endogenous Development Nucleus "19 de abril". In BoKvar City, BoKvar State, Venezuela.

Author: Amanda R. Olivier H.

SUMMARY

Conventional conventional impacts negatively the environment, especially natural resources. The agricultural production system of the Endogenous Development Nucleus "19 de abril", has been developing conventional agro-productive activities, which have generated an impact on natural resources, through practices such as deforestation, use of agricultural machinery, diversion of water bodies, chemical control of pests and diseases, without conservation of plant biodiversity and other natural resources such as water and soil, without observing any productive increase. To achieve the objective of designing the proposal of an integrated system of agricultural production as an agroecological alternative in savanna ecosystems in the Nucleo de Desarrollo Endogeno 19 de abril, the research methodology is of a projective or feasible and descriptive project type, with a field and documentary design, using participant observation, questionnaires and interviews. The population sample was represented by seven producers working within the Endogenous Development Nucleus "19 de abril". Upon analyzing the results, it was determined that it is necessary to promote a new agricultural model with the participation of the collective in all its dimensions, which contributes to the development of sustainable agriculture and conservation of natural resources based on agroecological principles. The design of a proposal for an integrated system of agricultural production as an agroecological alternative in savanna ecosystems in the Endogenous Development Nucleus "19 de abril" is proposed.

Key words: Agroecology, Integrated systems, Sustainable agriculture.

INTRODUCTION

The conventional approach to agriculture that was once valued for increasing productivity has produced increasing productivity has produced increases important in agricultural productivity agricultural productivity achieved a coverage significant increase in food supply. However, despite these achievements, this model has been affecting the environment, especially natural resources such as forests, soil, water, plant and animal biodiversity and human health.

In the last two decades, different agricultural development theories have emerged, generating proposals aimed at seeking a harmonious relationship between agriculture and the environment, with Agroecology standing out, whose main objective is the optimization of the agroecosystem or integrated production system.

Land degradation is rooted in economic, social and cultural factors and in the fragmented or atomized vision, which result in the over-exploitation of resources and inadequate soil and water management practices, leading to the loss of soil and consequently of its productivity, causing a reduction in the yield of agricultural production, which affects the quality of life of current and future generations.

The timely availability of reliable and meaningful information on the potential and limitations of natural resources is a prerequisite for integrated resource planning and management. In particular, information on land use options is of vital importance in order to plan for sustainable agricultural development and thereby achieve conservation of natural resources.

Local, regional, national and global food security requires a deep knowledge of productive agroecosystems. The Venezuelan State has been developing policies and programs related to the nature of the productive process, with the social, cultural and technological aspects that farmers have traditionally practiced throughout history; however, it is necessary to implement training and evaluation plans directed towards the institutions that are responsible for agricultural development.

This is why the conception of integrated agricultural production systems is strategic, aiming at the integration of social, economic, cultural and legal aspects, as sustainable alternatives for food production that lead to the development of a new agriculture that is more environmentally friendly, ecologically viable and socially just, sustainable agriculture.

The agroecological approach to integrated agricultural production systems makes it possible to evaluate the knowledge and skills of farmers, identifying the potential they have derived from centuries of applying traditional techniques to overcome the limitations of their productive agroecosystems and meet production objectives.

The objective of this research is to design a proposal for an integrated system of agricultural production as an agroecological alternative in savanna ecosystems and at the same time to function as a space for learning and integration between the different components that make it up as well as the valuation and enhancement of human capabilities, and the optimization, use, defense and protection of internal resources that are available to the production unit.

This is why it is important to propose alternatives to improve agricultural activity in harmony with nature, where families, communities or organizations adjust to their reality, either to correct their own weaknesses or to enhance their virtues and / or wealth, where a series of tools are put into practice that always tend to balance and complementarity between production and protection.

It is important to emphasize the need for an agricultural development model that protects natural resources, guarantees agricultural production and sustainable agriculture through the implementation of alternatives that maintain the balance and interrelation of each of the components of the production systems, contributing to the food sovereignty of our peoples.

The methodology used was descriptive and projective or feasible project, using as design the field and documentary research; the instruments and techniques of data collection were the observation guide, the survey, the interview and the questionnaire, for the physical and natural characterization in a participatory way of the study area, the description of the economic, social and technical-productive aspects of the Nucleo de Desarrollo Endogeno "19 de abril", as well as to know the economic, social and technical-productive aspects of the Nucleo de Desarrollo Endogeno "19 de abril", as well as to know the conditions of the area. Conditions agro-productive conditions of the same.

This research work aims fundamentally to transform a concrete diagnostic reality in the agricultural production system of NUDE "19 de abril" and thus contribute with the design of a proposal for an integrated system of agricultural production as an agroecological alternative in savanna ecosystems to protect natural resources and develop a sustainable agricultural model and in turn contribute to sovereignty and food security.

The development of the research work was structured in VI chapters. Chapter I presents the problem statement, the objectives of the research, justification of the study, characterization and advantages of the study area. Chapter II includes the background of the research, theoretical references related to agriculture and environmental degradation, and the new vision towards a renewed approach to agriculture. renewed approach to agriculture, the agroecology, the legal basis where the the agricultural legal framework and the
and some concepts that support the proposal.

Continuing with Chapter III, it refers to the methodology of the investigation indicating the type and design of the investigation, also, it includes the instruments and techniques used for the collection of data, the operationalization of the variables, the reliability and validity of the instruments used in the diagnosis of the same. Chapter IV contemplates the analysis and interpretation of the research results, the different indicators evaluated for the characterization of the study area and their analysis through tables and bar graphs, and finally, Chapter V corresponds to the conclusions of the research and Chapter VI, the formulation and design of the proposal. As a final aspect, a list of bibliographical references is made.

CHAPTER 1

PROBLEM

1.1 PROBLEM STATEMENT

The country's agro-productive development under the conventional agricultural production model or the "green revolution" has become increasingly unproductive and dependent on external resources; Numerous diagnoses assert a great deterioration of the different ecosystems and their natural resources caused by the implementation of conventional agricultural production techniques (Comerma and Paredes, 1998), increasing today the need to look for productive development strategies that articulate economic growth with social equity and that, in turn, do not degrade the productive potential of these plant formations, where man and nature can coexist in perpetual and harmonious unity.

Research approaches and development projects in the agricultural and livestock sector have been of a piecemeal nature with technologies that totally depredate natural resources, in addition to their exclusive focus on economic profitability, without taking into account the environmental balance and the needs of the community. According to Centeno (2000), 40% of the natural forests in Venezuela have been subject to agricultural development, being highly intervened by man in search of expanding agricultural frontiers without considering fundamental aspects that contribute to preserving the environment and satisfying human needs.

The approach to the development of agricultural projects under a conventional agricultural model, totally isolated from the reality and needs of the communities, has been generating a series of environmental and socio-productive affectations, as evidenced in the Endogenous Development Nucleus "19 de abril", where since 2005 an agro-productive project has been developed in an area of savannah ecosystems, conventional agricultural practices have been implemented, affecting the natural resources that characterize this type of vegetation formation, mainly generated by deforestation, uncontrolled use of agricultural machinery, diversion of water bodies, chemical control of pests and diseases, without establishing conservation techniques of plant and animal species and without observing any productive increase and therefore improvement in the living conditions of agricultural producers.

The current deterioration of savannas and other natural areas for the purpose of expanding agricultural frontiers has been a central topic of discussion by organizations such as the Food and Agriculture Organization of the United Nations (FAO)[1] , which states that mechanisms should be promoted to strengthen a new agricultural development model that conserves land, water, plant and animal genetic resources that do not degrade the environment, that is economically viable and socially acceptable, and that guarantees food security, equity, social stability and the participation of the population, water, plant and animal genetic resources that do not degrade the environment, that are economically viable and socially acceptable, and that guarantee food security, equity, social stability and the participation of the population in the general development process.

Nunez (2002) states that in order to promote the new productive model of agricultural development in the country, it is necessary to study alternative, strategic agriculture or paradigm changes such as agroecology, which has

The study is based on integrated agricultural production systems, which seeks an approximation of the guidelines of what is and should be a model of sustainable agricultural development, where the environmental, social, economic, cultural, technical, productive and historical aspects prevailing in our environment are considered and environmental risks are reduced, while maintaining the productive base of the productive base of our environment. sustainable agricultural development model, where the environmental, social, economic, cultural, technical, productive and historical aspects prevailing in our environment are considered and environmental risks are reduced, maintaining the productive base of agriculture over time.

Agroecology allows to analyze and organize a new model of agriculture and rural development. One of its important contributions is that it allows the elaboration of proposals for collective social action to challenge the predatory logic of the capitalist agro-productive model and replace it with one that aims at a socially just and ecologically sustainable agriculture (Gliessman, 2002).

Agroecology^a includes the dimension of the deterioration of natural resources in the in agrarian systems and proposes a new rationality and effectiveness to the different techniques and local actions , contributing to the dismantling of thedismantling agricultural models

---

[1] Land evaluation for irrigated agriculture. Rome Conference. 1999.

conventional, for its transforming action of the techniques and practices differentiated in a new theon of production (Sevilla, 2001).

It should be noted that since 2006, the Ministry of Popular Power for the Popular Economy (MINEP), through the office for the follow-up and evaluation of public policies for the socio-economic development of the nation and the acceleration of a new productive model, as part of the effort for the creation of a new economic system, began to promote the Endogenous Development Nuclei (NUDE), providing support for the development of agricultural potentialities in their regions (Official Gazette 38.370 of Wednesday, February 1, 2006).

The Endogenous Development Nucleus (NUDE) "19 de abril", was established in the peasant settlement of the same name, in the Heres Municipality, under the figure of a cooperative social organization called Kamaropa 9898, created through the Vuelvan Caras mission, initiating the development of an agro-productive project with a focus on conventional agriculture, putting into practice a series of agricultural techniques, in an area of 29 hectares of savannah ecosystems, considered as vegetal formations of great environmental susceptibility due to their ecological characteristics, altering the natural balance of this vegetal formation due to the effect of anthropic activities.

Consequently, it is intended to replan,considering participation of all those actors involved in the processes of agricultural development, with the aim of replacing conventional agriculture with a conservationist agriculture, in the same way, this work proposal will allow a significant contribution to the planning processes of agro-productive development of the country.

For the design of an agricultural development project, one must systematically consider the economic, social, environmental and technical-productive characteristics, as well as the physical and natural aspects, fundamental pillars of sustainability, which should guide changes that contribute to the true development of a more humane agriculture.:

Has the conventional agricultural development model had an impact on the deterioration of the savanna ecosystem present in the Endogenous Development Nucleus "19 de abril"?

Were the producers involved in the decision making process for the development of the agricultural project in this crop formation?

Does the savanna ecosystem meet the agro-productive conditions to carry out a conventional agricultural development project?

1.2 RESEARCH OBJECTIVES

1.2.1 General Objective

Design the proposal of an integrated system of agricultural production in savanna ecosystems as an agroecological alternative to promote sustainable agriculture in the Endogenous Development Nucleus 19 de abril, Parroquia Jose Antonio Paez Municipality" Heres of the state of BoKvar.

1.2.2 Specific objectives:

To characterize in a participatory manner the physical and natural part of the production system agncoladel Development Nucleus Endogen "April 19".

Describe the economic, social and productive technical aspects of the agricultural production system of the Nucleo de Desarrollo Development Nucleus " 19de April".

To know the agro-productive conditions of the agricultural production system of the Nucleo de Desarrollo Endogeno "19 de abril".

1.3 JUSTIFICATION FOR THE STUDY

In this order of ideas, this research proposes to diagnose and demonstrate the agro-productive conditions in the Nucleo de Desarrollo Endogeno 19 de Abril, in the savanna ecosystem, and proposes its transformation through the design of an integrated system of agricultural production, as an agro-ecological alternative that allows preserving this vegetal formation and the natural resources present and contributes to an agricultural development articulated to conservationist processes and to the production of environmental goods and services, conservation of local species strongly affected by the conventional agricultural activity, which satisfies essential social needs that go beyond the mere biological survival of the human species and at the same time, solve the productive needs in the agricultural area of the region.

In order to contribute to a successful formulation of the design of a proposal for an integrated system of agricultural production as an agroecological alternative, the aim is to reorient the practices that are being developed, where the scientific-technical part is exchanged or combined with the ancestral knowledge and promote or encourage new attitudes in the participatory collective and in turn develop a new integrated concept of natural ecosystem man, thus improving the level of quality of life and help stop the deterioration of natural resources; which generates positive consequences and effects on productive activities.

Knowing the agro-productive conditions and the economic, environmental, social and technical-productive aspects of the Endogenous Development Nucleus "19 de abril", will allow the implementation of the agro-ecological approach in accordance with the needs of the environment, as an innovative and holistic science that shows the alternative management of agriculture and helps in the transition of sustainable agricultural practices.

1.3 CHARACTERIZATION OF THE STUDY AREA

1.3.1 LOCATION

The NUDE "19 de abril", is located in the 19 de abril peasant settlement, in the Jose Antonio Paez parish, Heres municipality, at kilometer 11 of the highway that leads from Ciudad BoKvar-Puerto Ordaz, covering an area of 29 hectares of savanna ecosystem with the presence of a galena forest and a natural water body, with sandy loam soils and a relatively flat topography, with an elevation of 78 meters above sea level.

The zone where the study area is located is characterized by a dry period from January to June and a rainy period from July to December, with an average annual rainfall of 800 mm and an average temperature of 29°C, according to the meteorological station of the Bolivarian military aviation of Ciudad BoKvar. According to Comerma and Paredes (1998), the soils are classified as *Typic Haplustalf,* very susceptible to erosion and highly stony.

The access to the agricultural production system of the Nucleo de Desarrollo Endogeno 19 de abril, can be made through the main entrance to the 19 de abril peasant settlement, which is located at kilometer 9 of the Ciudad BoHvar Puerto Ordaz highway, it can be reached through an asphalt road in an area of five kilometers and then by the same road three kilometers of dirt road.

NUDE April 19 was initiated with the participation of 14 associates, eleven women and three men, who were organized under the figure of a cooperative, remaining at present seven associates. It should be noted that agricultural activity is part of the daily life and life strategy of this collective, it should be noted that this area has never been cultivated or intervened. It was in 2004 when the process of land movement and deforestation began in order to implement an agricultural development project under a conventional approach.

Figure 1 shows the relative national, regional and local situation of the study unit, and some points of interest where agricultural activities have been developed, such as the planting and cultivation of some crops like cassava (*Manihot esculentum*), fruit trees such as lechosa (*Carica papaya)* and parchita (*Passiflora edulis*), musaceae and the construction of infrastructures to support production such as broiler hatcheries and the Australian tank.T.M. coordinates of the geographic location of the study unit in the REGVEN system, determined through a geographic positioning system or GPS eXplorist500 type

Figure 1 Situation of the NUDE agricultural production system April 19.
Source: Elaboration INTI BoKvar 2008

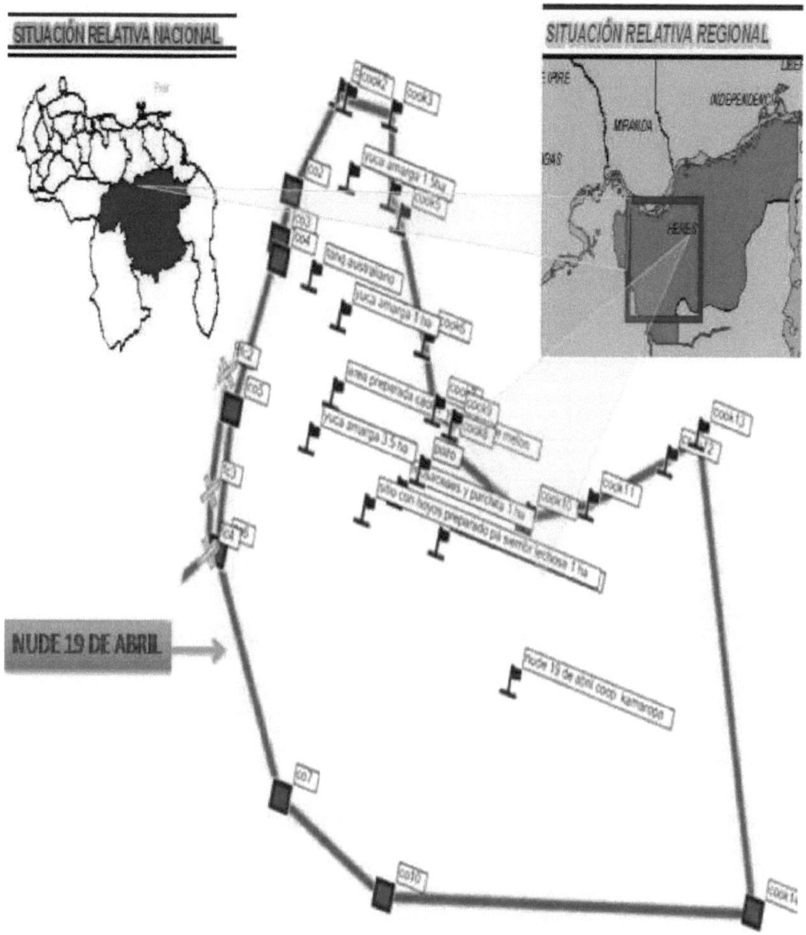

Leites of the study unit.

North: Marhuanta River.

South: Ciudad BoHvar-Puerto Ordaz Highway.

East: Caseno" Chaparral."

West: Vacant land.

Table 1. U.T. M Coordinates Geographic location of the agricultural production system of the NUDE April 19.

Points	Coordinates U.T. M	
	North	This
Co 1	309151	819575
Co 2	308921	820539

Co 3	309480	820602
Co 4	309382	820823
Co 5	309360	820856
Co 6	309661	820872
Co 7	309898	821296
Co 8	309362	820966
Co 9	309564	821272
Co 10	309868	823523
Co 11	309689	824256
Co 12	309866	826589
Co 13	309989	828162
Co 14	319789	826972

Source: INTI Bolivar 2008

Advantages of the study unit for the realization of the investigation:

- The local reality, its potential, problems and needs are known.
- We know the background of the agro-productive development of the agricultural production system.
- It is easily accessible.
- The willingness of some agricultural producers to support and participate in the research.

Other elements for a comprehensive characterization are included in Chapter III.

CHAPTER II

THEORETICAL FRAMEWORK

2.1 . Background of the investigation

The background information refers to previous research that has been carried out on the subject under study. In this sense, the following works were reviewed.

Chauran, Nieves and Gomez (2001) researchers of the Department of Agricultural Economics and Social Sciences, School of Agronomical Engineering, of the Universidad de Oriente Nucleo Monagas, identified the agroecological, socioeconomical and technical productive components and their interrelation and influence among them, in their research work titled the agricultural production systems at local level, their interrelations and influences among them, to achieve the objective of characterizing the conventional systems of agricultural production in the locality of La Guanota, Caripe municipality of Monagas state, using a descriptive and field research, for which data were collected through a survey, applied to the entire population, which was represented by 39 producers.

A general diagnosis of the area was made and the agricultural activity developed was described, the social aspects of the producers were considered, as well as the agronomical, environmental, technological and economic aspects of the crops involved, the results obtained showed that the production systems are formed by the cultivation of onion, chives and other crops.

(*Allium* schaenoprasum), Spanish Celery (*Apium graveolens*, var Dulce L), Lettuce (*Lactuca sativa*), Cilantro *(Coriandrum sativum* L), among others, 89.7% of the producers surveyed are male, agronomic work is done manually, 94.9% apply fertilizers and/or pesticides to their crops, 87.2% of the producers sell their products to truck drivers, they maintain that the level of technology in the area is very low and in some cases rudimentary, it was determined that the problem exists and it was suggested that the level of technology in the area is very low and in some cases rudimentary,2% of the producers sell their products to truck drivers, they maintain that the level of technology in the area is very low and in some cases rudimentary, the existing problem was determined and the strategy of using integrated agricultural production systems was suggested as an alternative to increase the quality of life of these producers.

The contribution of this work to the present research is that it facilitated the understanding of the diagnosis of the production unit in the environmental, social and technical aspects of production, as fundamental elements that represent the object of study.

Sarmiento, 2007 in his institutional work entitled conservation and agro-environmental production of the production systems in the zone of influence of the Guavio, with the objective of characterizing the agro-productive systems of the rural zone of the Guavio in the department of Cundinamarca in Colombia within the framework of the Natural Heritage Agreement - CORPOGUAIVO-EMGESA, where 34 surveys were conducted to producers located in the Round of the reservoir in the municipalities of Gachala, Gama and Ubala, in the process of gathering information it was necessary to go to dynamic producers, propositive and collaborators, the system approach was used as a methodology, which provided information of all the components of the agro-productive systems, the current existing interactions, the imbalances and the imbalances of the agricultural production systems. the existing interactions, imbalances and potentials.

To facilitate the study and understanding of agricultural activities, the different production systems were divided into three systems: livestock, mixed and fruit production systems. The results showed that integrated production systems allow the organization and management of agricultural practices with preservation of the environment, concluding that integrated agricultural production systems allow the combination of a wide range of activities, production items and objectives because they allow the greatest number of interactions between the number of components of the identified system.

The contributions of this research are that the revision of this work contributed to have an approach to the reality of the Nucleo de Desarrollo Endogeno "19 de abril" through the systematic approach as well as the criterion of division of the study area to facilitate the study and understanding of the same.

In this order of ideas, the interdisciplinary work team of research on livestock, forestry and aquaculture production systems, recognized, consolidated and linked with the society and national institutions, which proposes solutions to the demands on the improvement of the productive systems of the country of Mexico, led by Dr. Clemente Fernando in 2008, developed a research work entitled the agricultural production systems, forestry, aquaculture and wildlife for recreation, maintenance and sustainability of natural resources in the town of San Luis Potosi, forest, aquaculture and wildlife production systems for recreation, maintenance of the sustainability of natural resources in the town of San Luis de PotosL whose objective is to generate information for the establishment and operation of sustainable and integrated production systems as alternatives to increase the level of production, improve the standard of living of producers and society in general. society in general.

Under a field methodology, demonstration plots were implemented in 12 production units with the participation of 42 producers and 5 researchers where the productive potential of alternative crops were evaluated in accordance with the quality of rainfed irrigation water and clean, green and ethical methods were developed to improve productive efficiency, the results obtained allowed the generation of alternatives to improve food efficiency, establishing integrated agricultural production systems as alternative production structures, since agricultural development requires a new approach to meet food needs that are socially just, ecologically and economically sustainable and, above all, culturally acceptable.

The contributions of this work for the present investigation is that for the study of agricultural production alternatives, participatory research tools and qualitative research methods are used, such as interviews, diagnostics, and systemic analysis, since it helps to place the participants at the center of the development and helps to fulfill its objectives, reflecting the importance of community participation for the establishment of any agricultural development proposal, as well as the need to diagnose, characterize and understand all aspects of the production units to determine their potential and resources, as well as systemic analysis to understand the various components and their interactions necessary for the establishment of an integrated agricultural production system.

La Roche (2006). In her work entitled Characterization of livelihood strategies, technologies and agricultural production practices of rural households in Esparza, Costa Rica, the objective of which was to characterize the livelihood strategies and agricultural production technologies of rural households in Esparza, Costa Rica, was framed within the integrated agro-productive approach to ecosystem management project. The livelihoods approach was used to identify and analyze the capital assets of rural households. Tools used to obtain information included questionnaires, participatory observation, key informant interviews and triangulation in rural households in a tropical forest area.

The results reflected homogeneity in the activities of the activities agricultural production without any type of diversification, both for self-consumption and market-oriented, without any type of orientation or technical advice, do not belong to any social organization, the rural household members are older than 40 years, which indicates a certain aging of rural households, which affects the dedication and interest in agricultural activities, The use and maintenance of agricultural production systems, concluding the importance of knowing the life strategies of the communities and the socio-productive elements that contribute to recover degraded soils, increase the biodiversity of plant and animal species and other natural resources and undoubtedly improve income, in order to establish environmentally friendly production systems.

Within this order of ideas, the contribution of this research work to the study reflects the importance of knowing the agro-productive practices and their effect on the environment, as well as knowing the diverse life strategies of the rural households existing in different areas of the country. At different at different times of the year year.

In Venezuela, the training received in schools and universities from 1970 onwards has not always prepared professionals in the agricultural sector to understand the farmer's rationality, emphasizing the application of concepts and techniques of the green revolution, preventing the understanding of the holistic approach and hindering the formulation of concepts and instruments adapted to the farmer's realities.

In Venezuela, the agroecological model has been advancing since 2004, developing educational and productive technical programs, responding to the strategies of a socialist productive model.
established in the National Plan of economic and social development 20072013, with the creation of the formation program in agroecology, the Bolivarian University of Venezuela, has trained 156 professionals in agroecological production at the technical level[2] which in its majority are inserted in the institutions of the State, responsible for agricultural development as is the case of the Foundation for Training and Innovation to Support the Agrarian Revolution (Ciara) who have been promoting some agricultural projects under the agroecological approach in some states of the country such as BoHvar, Zulia and Lara.

2.2 THEORETICAL REFERENCES

2.2.1 Conventional agriculture and environmental degradation

Venezuela is one of the eight (8) richest countries in the world in biodiversity, the

[2] Statistics of the Agroecology Program of the UBV at the national level. 2009

greatest proportion of that variety of plants and animals is found in the natural forests of the country. The destruction and degradation of forests in Venezuela to expand agricultural land has become a threat to the ecological stability of ecosystems, causing a decrease in soil organic matter and biodiversity.

In the last years of the 20th century, about 6 million hectares of forests have been lost in Venezuela, mainly in the states of Barinas, Portuguesa, Monagas and BoHvar. According to Centeno (2000), the largest proportion is due to the conversion of forests to cropland or the development of agricultural projects under a conventional agricultural approach; however, there are no reliable statistics on forest destruction in Venezuela.

The expansion of agricultural frontiers in Venezuela maintains a continuous threat to the remaining forests, due to the tendency to increase food production mainly through agriculture.
expansion of agricultural land, instead of relying more and more on higher productivity per unit area on existing farmland or through the implementation of natural resource conservation and protection techniques applied to agricultural development (Nunez, 1999).

In the decades of the 70's, 80's and 90's, where the model known as "green revolution", which is the basis of conventional agriculture, was given preference, it has left a pernicious balance since the policies assumed by the institutions or organizations responsible for agricultural development, did not allow reducing rural poverty and hunger, especially in the countries known as third world countries, According to Casado (1997), the agricultural development techniques were not adapted to the needs and potential of the farmers because of the technological implementation based on the analytical parcel approach, breaking with the organicist vision of nature as a living entity, considering the land as a machine and the separation of the agricultural process from its connection with the other elements of the ecosystem.

Gliessman (2002) presents a review of some of the elements that conventional agriculture has generated in the environment, some examples are: intensive tillage has caused in Asia and Latin America the loss of 5-10 t/ha/year due to wind and water erosion and in Africa it has reached 30 t/ha/year. Conventional agriculture has proven to be unsustainable, due to the high costs of its inputs, technological and production problems, and the environmental, social and cultural problems caused (Leyva and Pohlan, 2000).

Conventional agriculture in forests involves the simplification of biodiversity, alteration of the water cycle and soil erosion by the development of monocultures that make crops vulnerable to pests and diseases, resulting in an artificial production that requires constant human intervention. In most cases, this intervention occurs in the form of agrochemical inputs, which, in addition to increasing yields, result in a number of undesirable environmental and social costs (Altieri, 1997).

With the advance of agricultural modernization agncola, the principles agroecological systems are continually underestimated. As a consequence, modern agroecosystems are unstable and their breakdowns are manifested as recurrent pest outbreaks in many cropping systems and in the production of food.

also in the form of salinization, soil erosion, water pollution and forest deterioration etc.

According to Altieri and Letourneau (1994), the worsening of most pest problems has been experimentally linked to the expansion of monocultures at the expense of plant diversity, which often provides key ecological services to ensure crop protection.

Many researchers, including Altieri, Restrepo, Rosset and Gliessman, have shown that although the "green revolution" increased productivity for a small number of farmers with large tracts of land and access to credit, the consequences for the environment and for the income of the vast majority of small farmers, The consequences for the environment and for the income of the vast majority of small farmers were adverse, because the use of high-yield seeds displaced a greater number of traditional varieties, eroding the biodiversity of crops, requiring the use of large doses of inorganic fertilizers and pesticides, causing chemical contamination of soil, water and human beings.

These declining yields and rising input costs are factors that led the Food and Agriculture Organization (FAO. 1998), at the end of the 20th century, to conclude that a new approach to agriculture is needed from a social and ecological point of view.

Agriculture concentrates a part of the population with the greatest inequalities, and because it is the source of a large part of the production of the of the production of foodstuffs, contribute significantly to the destabilization of the environment, the reduction of the productive capacity of soils due to erosion and desertification The destabilization of the environment, the reduction of the productive capacity of soils due to erosion and desertification, the loss of biodiversity, among others. Conventional agricultural activities and those related to the supply of inputs, as well as other activities of the rural population, are also having an impact on this growing degradation of the environment, mainly on the elements that make up the ecosystem, such as natural forests (Restrepo, 2000).

Actions such as the burning of forests in order to exploit new agricultural land, as well as the burning of savannas or crop residues, the use of pesticides and the application of nitrogen fertilizers have been increasing, thus contributing 17% of the world's carbon dioxide emissions and, therefore, to environmental pollution (Casado, 2004).

The accelerated destruction of ecosystems to develop agricultural areas is bringing serious problems in soil erosion, the acceleration of desertification processes, salinization and loss of biodiversity, which alters the conversion of wind systems and rainfall patterns, and the accumulation of carbon dioxide in the atmosphere will increase due to the effect of forest burning (Kaimowitz, 1997).

Casado (1999) points out that the loss of biodiversity is another important impact generated by conventional agricultural activity in natural areas such as savanna ecosystems, due to the displacement of local varieties by high-yielding varieties, which have been and are developed by research centers and which generally require high levels of chemical inputs to achieve their potential.

In order to build agricultural fields or develop agricultural projects, the destruction of tree lines has been generated in the boundaries of galena forests as well as in other places, opening spaces for the agricultural machinery required for this type of development of conventional agriculture to work, thus accelerating the processes of erosion and reducing the infiltration of water into the soil and also the biodiversity, especially of birds and insects predators of crop pests, as well as eliminating through diffusion and migration a part of the beneficial fauna and biodiversity (Suquilanda 1999).(Suquilanda 1999).

It is necessary to build an agro-productive model based on solidarity, community life and more humane, able to accept and combine with popular knowledge, which helps to spread a social, ecological and solidarity conscience with nature, to ensure the preservation of future generations, putting into practice new poKtical, technological, social, cultural and educational proposals, which contribute to stimulate a new ethic of production based on a humanistic naturalism, agroecology (Rodriguez, 2007).

2.2.2 The return to organic farming: A new vision

Since the 1960s, when the decline of agricultural production processes began, driven by the scientific-technical conceptions of the "green revolution", different theories aimed at seeking harmony between agriculture and the environment have re-emerged. The development of a series of agricultural practices aimed at alleviating the various problems caused by conventional agriculture, affecting, in general terms, the organization of production, its social base and especially cultural and ecological values, ignoring for a long time the importance of environmental factors for the functioning and maintenance of the economic and social system. (Garda, 2004).

In the face of a world dominated by superficial life, alienated, based on the market as a parameter of planetary measurement, individualism, overproduction, consumption and the mercantile use of science and technology, we must impose a world based on a human and environmental logic, which takes into account all the interrelationships between society and nature. In function of this, it is necessary to rethink the whole conventional agricultural production system, which, as a model, is in clear decadence, and to promote one that breaks with local, national and international structures, more in accordance with the reality of our times (Guzman, 2000).

The statistics and daily life reflect an approximation of the daily facts, observing a worrisome approach to the critical levels of this problem, while the inefficiency of the political measures adopted historically is evident, so it is considered pertinent to approach it from an integral conception of the causality of the problem and its foundations, among which stand out rural development, based on food production and the rural component of the consumption subsystem, thus corresponding to the determinants of the initial and terminal phases of the food security system.

Motivated by the mechanistic and reductionist approach of the green revolution model that has so far prevailed and has been used mainly in agricultural development projects, it has proven to be inadequate to solve the complexities of agroecosystems and is being questioned by different sectors, In view of this, there is a need for a different approach to conventional agricultural development, based on a paradigm that is more responsive to human needs, that considers social and ecological systems, as well as an approach to agriculture that is more environmentally and socially sensitive, focused not only on production but also on the ecological sustainability of the agricultural system.

In view of the above, efforts must be made to interpret the reasons for a new agriculture, the scientific principle of which is agroecology, whose approach is based on a holistic and multidisciplinary approach that unifies the socioeconomic and technical perspectives with the design, management and evolution of the production system with its existing social and cultural base, implementing the use of low-input techniques and giving rise to a new participatory epistemology with a poKtical character (Funtowic, 1994).

According to Gliessman, (2002), agroecology is emerging today as the fundamental science to guide the conversion from conventional production systems (monocultures dependent on agrochemical inputs) to more diversified and self-sufficient systems, agroecology uses ecological principles that favor natural processes and biological interactions that optimize synergies in such a way that agro-biodiversity is able to subsidize key processes such as organic matter accumulation, soil fertility and crop productivity.

In this sense, agroecology is a different approach because it is an integrative scientific paradigm, which is part of an alternative lifestyle that rescues and enhances socio-environmental and ancestral knowledge in an intercultural, transdisciplinary and democratic dialogue with science. Here the social and agroecological systems are reflected, in the sense of promoting and developing transition processes that make agricultural activity more sustainable by producing healthy food (Altieri, 2000).

It follows the relevance of moving towards agricultural practices that make agriculture based on technical applications which must be ecologically appropriate to the preservation of agro-ecosystems in order to favor those management options appropriate to the objectives of a truly conservationist agriculture.

Agroecology is primarily the study of agroecosystems or integrated production systems, considered as the result of a co-evolutionary process between society and nature, it is also, and consequently, a way of producing food counting with nature and not against it, inserting itself in the territory through appropriate technologies (autochthonous varieties and practices that protect the ecosystem as a whole), as Gliessman (2002) points out.

This science, thus understood, starts from a principle of austerity in the use of inputs, especially energy, and relies on a popular and collective knowledge rooted in peasant and indigenous wisdom and rationality that capitalist modernization has banished (Restrepo, 2000).

The application of the same in the ecological management of forests and natural resources through forms of social and collective action presents alternatives to the current crisis of agriculture, since its systematic nature, considers the production unit, the community organization where the local knowledge systems are found, peasants bearers of the endogenous potential that allows enhancing the ecological and socio-cultural biodiversity (Guzman, 2000).

In order to generate a reflexive thinking in the communities, mainly those of the rural sector that leads to a true food self-sufficiency, it is essential to consider the idea of incorporating conservationist technologies and social organization as a way to develop the capacity of a permanent productive activity, significantly valued by agroecology.

Accordingly, Guzman (2000) states that agroecology is based on an epistemological assumption that proposes a break with the conventional paradigms of official science, claiming the necessary unity between the different natural sciences and social sciences such as anthropology, sociology and ecology to understand the interconnection between ecological, economic and social processes and the essential link between soil, plant, animal and human beings.

This perspective leads to the development of integrated agricultural production systems that seek to achieve an efficient utilization of the farm's own resources, and to maximize the relationships between the components of the system in order to improve the biological basis, economic and technical viability.

As Altieri (1999) argues, food can be produced in accordance with the diverse agroecosystems, making use of the environmental reality of the producer and the mechanisms that nature employs in its own development, as contemplated by the agroecological approach, since it is based on broader philosophical premises than conventional agriculture. This paradigm illustrates that a social and environmental relationship must be established, giving legitimacy to the farmers' knowledge in order to solve the problems of rural poverty and environmental deterioration.

It is necessary to establish self-sufficient and diversified agricultural systems of low input utilization, as a life project for the peasant families that, besides assuring an abundant and rich food in proteins, vitamins and minerals, helps each one of its members to live in harmony with nature, preserving and enjoying the environment that surrounds them, breathing clean air, avoiding the felling of forests, conserving water bodies and promoting the improvement of land and crops, thus guaranteeing the production of food in the present and for future generations.

2.2.3 Agroecological practice in Venezuela

In Venezuela since 2004, the Instituto para la Production e Investigacion de la Agricultura Tropical (IPIAT), a non-governmental organization that supports the transition towards agroecology, reports the development of agroecological practices in 74 productive communities in seven states of the country, mainly located in the western region, where more than 100 different crops are produced under this model (Nunez, 2007).

Agroecology is revolutionizing the Venezuelan agricultural collapse, since a good number of peasants and agricultural producers have adopted the path of transition towards the consolidation of sustainable agriculture as defined in article 305 of the Constitution of the Bolivarian Republic of Venezuela of 1999. Authors such as Nunez (2007), establish that the Venezuelan revolutionary process is gestating an unprecedented and authentic policy of agro-food sovereignty based on agroecology as a tool for social transformation, since at the present time at the level of many state institutions such as the Integra Rural Development Institute (INDER), the Social Agricultural Development Fund (Fondo de Desarrollo Agncola Social de Venezuela), the National Agricultural Development Fund (Fondo de Desarrollo Agncola Social de Venezuela) and the National Agricultural Development Institute (INDER), The Socialist Agricultural Development Fund (FONDAS) and others with responsibilities in agricultural development have been sensitized to initiate actions aimed at the consolidation of agroecological sciences and are implementing the implementation of agricultural development projects based on agroecology, Table 2 shows the government institutions and the main activities in support of agroecology.

Governmental Institutions and main activities in support of the Agroecology

Government Institutions	**Agro-ecological activities**
Ministry of Agriculture and Lands	Integral Agricultural Development Plan 2007-2008
InstituteNacionalde biofertilizersand73 Agricola Integral (INIA)	Health113 laboratories laboratories laboratories to cover between 2 and 3 million hectares in 2010.
National Institute for Rural Development (INDER)	Agroecological Schools

FONDAFA	24 billionfor finance14 crops agroecological and conucos.
National Institute of Agricultural Research (INIA)	Sinedes; various investigations
Foundation Training Innovationfor Support Revolution Agraria (CIARA)	Organic Coffee Program it
National Land Institute (INTI)	Initiatives in some Zamorano farms
Ministry of Higher Education	Bolivarian University of Venezuela (UBV), ILDEA
Ministry of Science and Technology	Various Investigations
Fundacite Merida	Learning Network at Agroecology; 52 projects.

Source: Nunez 2007.

The above table allows us to sustain the thesis of the advances in agroecology at the institutional level, except in some cases such as in the Valle de la Mesa de Guanipa, Anzoategui state, 4,000 hectares of soybean (*Glicine maximus)* are being developed as a monoculture and in the planting plans for 2009, 12,589 hectares of yellow corn (Zeas maize) and 22,477 hectares of white corn (Zeas maize) were planted as a monoculture in San Francisco de la Paragua, Anzoategui state, also as a monoculture.In 2009, 12,589 hectares of yellow corn (Zeas *maize)* and 22,477 hectares of white corn (*Zeas maize) were planted* as monoculture in San Francisco de la Paragua, BoKvar state, and according to the Integral Program for Agricultural Development (PIDA) presented to the Ministry of People's Power for Agriculture and Lands, 22,000 hectares of corn are planned to be planted in 2010.

It should be noted that each of these agricultural projects are accompanied by packages of agricultural inputs of chemical origin as well as the use of soil tillage which is carried out mechanically, showing the contradictions between a conventional model and a conventional model. conventional model questioned and questionable, and the agroecological model considered by the state as a strategy, but perhaps not on a significant scale.

In the municipality of Heres in the state of BoKvar there are some cases of the application of state agricultural policies under an agroecological approach in rural communities, such as Agua Blanca, Los Piquito and others, coordinated by institutions such as the Foundation for Training and Innovation to Support the Agrarian Revolution (Ciara), who have been supporting 23 rural and peri-urban communities providing training and support to independent producers, families and social organizations, serving approximately 183 people in 2007 and 154 people in 2008 (Ciara, 2009).

More recently, the Fondo para el Desarrollo Agrario Socialista region BoHvar (FONDAS) included in its credit portfolio the agro-ecological approach and has allocated 40 credits within this modality for agro-ecological plots with surfaces of 2 hectares located in the municipalities Heres and Caroni.

The Bolivarian University of Venezuela in its different branches in the states of Falcon, Zulia, BoHvar and the city of Caracas has been training university graduates and technicians since 2005; Some of them are part of the employees of the Fondo de Desarrollo Agrario Socialista (FONDAS), the CIARA Foundation, the Instituto de Sanidad Agncola Integral (INSAI), the Corporation Venezolana de Guayana (CVG) in the Agricultural Development Department, and the Instituto de Investigaciones Agncolas (INIA), which leads us to suppose that the agroecological approach is being supported by the policies and programs of the national government.

2.2.4 Integrated Agribusiness Production Systems (SIPA)

With the implementation of integrated agricultural production systems (ISAP), a change in the conventional approach to agriculture has been established, which, although it has produced important increases in agricultural productivity and has achieved significant coverage in food supply, has also produced indicators that describe the impact on the environment, especially on natural resources such as forests, soil, water and plant and animal biodiversity.

Rodnguez (2007), considers that integrated agricultural production systems, not only represent physical spaces or production models focused on the economic, social and social aspects of the economic, social and environmental environmental, at which may be exercised the art of agriculture in order to convert itinto a life project that satisfies the nutritional and economic needs, as well as the spiritual and emotional needs of the person, the family and the community.

The SIPAs seek to carry out a mode of agriculture that seeks to provide short, medium and long term production, through the use of production methods conducive to the stability of the production unit, the recycling of organic matter and nutrients, encouraging local food production, adapted to the socio-economic and natural setting, where the production unit is considered an ecosystem. According to Rodriguez (2007), agriculture should not be oriented only to the search for high yields of a particular product, but rather to the optimization of the system as a whole, thus promoting an integral productive agricultural system.

The most important characteristic of the integrated agricultural production systems with agroecological bases is that it seeks mainly to conserve nature, since it depends on it that the production systems are sustainable, including the conservation of soil, flora and fauna and water bodies. It also contributes to the conservation of organic matter, through its diversified production, to the balance of beneficial fauna, to the use of available resources, to the maximization of land use, to the reduction of economic risks and to a more balanced nutrition.

The SIPAs allow to retake locally the agricultural diversity in time and space with the use of agroecological practices and techniques such as crop rotation, polycultures, agroforestry systems, where it is possible to reduce production risks, as well as positively influencing biodiversity and increasing the stability of the agroecosystem.

The conception of sustainable agriculture through integrated agricultural production systems with agroecological bases, requires that the agricultural production unit be considered as a global ecosystem, where the optimization of resources is sought, through the use and implementation of agroecological fundamentals and principles applicable to the management of agroecosystems that allow obtaining biological sustainability and economic viability in the production system.

It is also important for the development of FFS as an objective basis for food security, to support local production of crops adapted to the natural and socio-economic environment, to sustain a desired net production through the preservation of natural resources, to reduce costs and to increase the efficiency and economic viability of small and medium producers, thus promoting a diverse and flexible agricultural system, with the purpose of satisfying the basic needs of rural populations.

2.3 Legal basis.

Article 299 of the Constitution of the Bolivarian Republic of Venezuela establishes that: "*The socio-economic regime of the Bolivarian Republic of Venezuela is based on the principles of social justice, democracy, efficiency, free competition, protection of the environment, productivity and solidarity*" (CRBV, 1999).

Venezuela has been developing a favorable legal framework for the protection of forests, through the inclusion of the Law of Forests and Forest Management, for the protection of natural resources essential for everyone, for its ecological, environmental, social and economic value, developing and adapting management plans as instruments aimed at the formation of a culture of preservation of ecosystems in the population, which leads to value and conserve our natural resources and native species, to the development of sustainable productive activities.

Article 127 of the CRBV establishes that: *"It is the right and duty of every generation to protect and maintain the environment for the benefit of itself and the future world"*.

Articles 128 and 129 state the obligation of the Venezuelan State to develop land-use planning policies and studies of activities that could degrade the environment and ecosystems, and it is from this approach that the SIPAs are proposed as an agro-ecological alternative for food production and to increase food sovereignty and consolidate food security under a focus on the protection of natural resources.

Article 305 of the Constitution of the Bolivarian Republic of Venezuela, when dealing with the socioeconomic system of the nation, establishes:

"That the state will promote agriculture as the strategic basis for integral rural development in order to guarantee the food security of the population, understood as the sufficient and stable availability of food at the national level and the timely and permanent access to it by the consuming public.... Food security will be achieved by developing and prioritizing domestic agricultural production. The production of food is of national interest and fundamental for the
Economic and social development of the nation".

This constitutional mandate corresponds to the provisions of the National Economic and Social Development Plan, stage 2007-2013, of the Simon BoKvar National Project, through a series of objectives, strategies and policies regarding food production, which establishes the New Socialist Production Model, one of its objectives being to increase food sovereignty and consolidate food security through the development of sustainable agriculture.

In view of this, Venezuela requires a food guarantee for a population of over 27 million inhabitants (National Institute of Statistics, 2008), whose organization and resulting dynamics, surely

will propose new spatial considerations, particularly in terms of production and population and spatial distribution of demand or consumption, the system must link production, distribution, marketing and consumption activities to achieve agrifood security.

The Food Sovereignty and Security Law approved on August 6, 2008, and in its article 4 establishes that : "The Food Sovereignty and Security Agrifood as the inalienable right of a nation to define and develop agrarian and food policies appropriate to its specific circumstances, based on local and national production, respecting national production, respecting the conservation of productive and cultural biodiversity, as well as the capacity for prioritized self-sufficiency, guaranteeing timely and sufficient access to food for the entire population".

And its article 5 also indicates that;". Agri-food security is the effective capacity of the State, in co-responsibility with the national agri-food sector, to guarantee to the entire population, the availability, access, exchange and equitable distribution of food in a stable manner, to ensure the physical and emotional conditions suitable for integral and sustainable human development, considering the exchange, complementarity and economic integration between peoples and nations as an essential element that guarantees the right to food".

These global guidelines oblige cooperation throughout the country of the large masses affected, especially the human groups with fewer resources and in more sensitive situations with respect to the levels of vital balance, and the public and private institutions with competence in the agricultural area, in the great task of developing initiatives that allow, through the understanding and management of the different magnitudes of the problem and its implications, to assume the responsibility of the application of pertinent measures that in their concretion and implementation, favor the control of the situation with the treatment of food sovereignty and security as an objective situation.

All these policies advanced by the Venezuelan State in productive matters are being supported by the participation and social control of local organizations such as land committees, cooperatives, social production enterprises, communal councils, community socialist production units, among other forms of organization as a means to guarantee the relevance and sustainability of development dynamics, as well as to favor the success of the new socialist productive associations, among other forms of organization as a means to guarantee the pertinence and sustainability of the development dynamics, as well as to favor the success of the new socialist productive associations, overcoming the economic and social inequalities of the different sectors considering the human being as a determining element for the achievement of a true agricultural development.

2.4 BASIC CONCEPTS UNDERLYING THE PROPOSAL

Agroecology: Science that is characterized by integrating the aspects economic, ecological, social and cultural aspects linked to agriculture, with the purpose of offering alternative management to those proposed by conventional agriculture, whose negative effects on the environment and the rural population are widely documented.

Agroecosystem: A system formed by a biotic community, which includes at least one agricultural population and the physical environment with which it interacts, processing inputs of energy and materials, which produce biomass outputs.

Integrated Agricultural Production Systems (IAPS): These are production units formed by a biotic community, which includes an agricultural population and the environment where the elements that integrate it interact for its perfect functioning.

Systems: An arrangement of physical components, a dynamic and interacting whole or a collection of things united or related in such a way that they form and/or act as a unit, an entity or a whole.

Savannas: Vegetation formations characterized by the presence of more or less scattered trees that exhibit herbaceous, arboreal and mixed associations, with soil strata dominated by grasses.

CHAPTER III
METHODOLOGICAL FRAMEWORK

3.1 Type of research

On the basis of Balestrini, 2006 and Garrido, 2007, it is a research defined as descriptive, because there is an interpretative and comprehensive understanding of reality, where the research scenario is seen in a real Hispanic perspective. According to Hurtado, (2008), it is an research projective or feasible project because proposes solutions to a proposed research. It establishes an alternative for change or solutions to a given situation, through the systems integrated production agncola as alternative agroecology in savanna ecosystems in the Endogenous Development Nucleus" April 19".

3.2 Research design

According to Balestrini, (2006), the information required must be sought exactly in the place of the facts or its natural environment, in this case in the agricultural production system of the Nucleo de Desarrollo Endogeno "19 de abril" and of documentary type, because it was resorted to record or documentary sources, in this case in the institutions or organizations of the State, linked to the agricultural development of the region such as the Foundation for Training and Innovation in Support of the Agrarian Revolution (Ciara) and the Fund for Socialist Agricultural Development (FONDAS). With respect to temporality, the design is retrospective evolutionary because it covers the period from 2004 to 2009, and with respect to the organization and extent of the data, it is multivariate of cases (Ram^ez, 2007).

3.3 POPULATION

A population or universe can refer to any set of elements of which we intend to investigate and know their characteristics or one of them and for which the conclusions obtained in the research will be valid (Balestrini, 2006).

The population sampled in the present research and to which we have access is conformed, from the qualitative point of view, by the 13 agricultural producers that conform the study unit, which is represented in this research by the agricultural production system of the Nucleo Desarrollo Endogeno "19 de abril" settled in an area of 29 hectares. Population is understood as the set of observational units that have certain common characteristics that homogenize them as a whole (Arias, 2006).

3.4 SAMPLE

The sample, as is known, is constructed with elements that make up a subset of the population, but this subset must be representative of the entire population, not of a part of it (Ram^ez, 2007), in this case it is of a non-probabiKstic type, which assumes, for the choice of the subset, arbitrary criteria on the part of the researcher (Balestrini, 2006).

In the present investigation, the sample is made up of 53.85% of the producers, which is equivalent to seven (7) agricultural producers, that is, four men and three women. It should be noted that this selection was intentional because these producers were the ones who had the greatest initiative and interest in working with integrated agricultural production systems with an agroecological approach, as they were also the pioneers in forming the social organization in which they work, develop and put into practice the different conventional agricultural activities in the NUDE, which is represented under the figure of an agricultural cooperative called Kamaropa.

At the time of their selection, some logical, important criteria and opinions were considered, such as: knowledge of agricultural production, subsistence activity, voluntary consent and a particular level of trust with the researcher.

3.5 NOMINAL DEFINITION AND OPERATIONALIZATION OF RESEARCH VARIABLES.

Lazarsfeld (cited by Ram^ez, 1999) defines a variable as a dimension of an object. Hollander (cited by Ram^ez, 1999) indicates that it is an attribute that can vary in one or more ways. In other words, a variable is an element that is broken down from the general to the particular, until it is brought to the most specific expression. This element can acquire different values.

They can be extracted from the general objective or the specifics and linked to the questions in the interview questionnaires or other instruments. instruments from collection instruments information.

3.5.1 Dependent Variable

Agricultural Production Systems: These are production units formed by a biotic community, which includes an agricultural population and the environment.

3.5.2 Independent Variable

Savanna ecosystem: Plant population with shrub species of the lenosa and graminiform type with high biodiversity content and biotic and abiotic resources, low fertility soils, with high infiltration and low moisture retention.

Table 3 shows the matrix of the operationalization of the variables with their dimensions, indicators and sub-indicators.

Table 3 Operationalization of the variables on the design of a proposal for an integrated agricultural production system as an agroecological alternative in savanna ecosystems in the Endogenous Development Nucleus "19 de Abril".

Dimension	Indicators	Sub indicator
Environmental	Vegetation	Type of vegetation
		Abundance of vegetation
		Predominant plant species
	Soil	Soil type
		Soil fertility
		Organic matter
		Texture
		Permeability
		Presence of macro nutrients (K, P)
		Presence of micronutrients (Ca, Mg and Na)
		Hydrogen potential
		Drainage
		Color
	Topograffa	Relief
		Pending
	Hydrograffa	Natural water body
		Artificial water body
	Weather	Temperature, precipitation and evaporation
Social	Social organization	Type of organization
	Land tenure.	Municipalities
		Private land
		INTI Lands
		Wasteland
	Social Services.	Utilities
	Agricultural credits.	Agricultural financing
		Institutions financing agricultural activities
	Participation.	Activate
		Passive
Production	Dive	Fruit trees

technician	rsityfrom crops.	Vegetables
		Rafces and tubercles
		Legumes
	Soil tillage	Mechanics
		Manual
	Pest and disease management.	Chemical pest control
		Ecological pest control

Continuation Table 3.

	Technical y assistance | training	Type of trainingand technical assistance
		Agricultural courses
Economica	Income level	Admission to system production NUDE April 19
		% of employment in NUDE's agricultural production system April 19
	Planting area	Cultivated hectares
	Deforestation	Deforested area
Agroproductiva	Agroproductive Project	Agro-productive activities and agricultural project implementation

Source: Own elaboration 2010

3.6 DATA COLLECTION INSTRUMENTS

In function of the objectives defined in the present research study where the design of a proposal for an integrated agricultural production system as agroecological alternatives in savanna ecosystems is proposed, a series of instruments were used to capture and register qualitative and quantitative evidence of the actions carried out by the Endogenous Development Nucleus "19 de abril". The information used comes from primary and secondary sources. However, since the initial research work consisted of obtaining secondary information from different sources, which later allowed more efficient collection of primary information during the field phase, their descriptions are presented in the following order:

Information from secondary sources.

- The information from secondary sources related to the study area was obtained through documentary analysis, from unpublished documents, records, and productive technical reports corresponding to the agricultural development of the CIARA Foundation, the Institute of Rural Development (INDER), the Ministry of People's Power for the Environment, and the Ministry of Agriculture.
Agriculture and Land, and the National Land Institute (INTI), from which information was obtained about the location of the study area, social, economic and technical aspects of production. For the documentary or bibliographic review, sources related to agroecology, agriculture in savannahs and forests and integrated agricultural production systems were used, as well as the analysis of theoretical content related to the characterization of integrated production systems as production alternatives.

in natural ecosystems. The information related to the soil was based on the soil analysis previously carried out at the Instituto de Investigaciones Agncola (INIA). Agricultural Research Institute (INIA
Anzoategui), as well as physical characteristics were determined through the construction of a test pit.

Information from primary sources.

- To characterize the natural physical aspects of the study area such as vegetation, soil and hydrography, and social, economic and technical aspects of production such as social organization, land tenure, income level, pest and disease management, crop management, we used the technique of direct observation and the instrument was an observation guide as presented in Annex 2, which is intended to complement and validate the information obtained in the interviews (Veliz 2009).

The information from primary sources was also obtained through questionnaires consisting mostly of closed questions and very few open questions, these questionnaires have the advantage that the analysis of the data is simple, provides the possibility of comparing and aggregating answers easily and the questions can be done in a short time (Veliz 2009). The questionnaire consisted of a series of questions asked to the seven cooperative members oriented to describe the environmental, social, economic and technical aspects of production such as social organization, social services, employment, land tenure as presented in Annex 3.

- Another primary source applied was the interview, a tool that can provide essential information for a better knowledge and understanding of a specific topic (Veliz 2009). Such interviews can be carried out with actors who are involved in the topic in question.

  Coordinator of the Ciara Foundation BoKvar branch and the coordinator of the Socialist Agricultural Development Fund (FONDAS) in order to obtain the following information information information on land tenure, forms of participation of the cooperative in training courses and to learn about the agro-productive situation of the production system of the Nucleo de Desarrollo Endogeno "19 de abril". The members of the cooperative were interviewed to obtain information related to the incidence of the productive activity on their family life, as shown in Annex 4.

- Different means were used to record the information, such as a field notebook and a photographic camera.

1.7 VALIDATION

Validation refers to the degree to which an instrument actually measures the variable it is intended to measure (Guion, 2002). The instruments applied were subjected to external evaluation for their validation of judgment by experts, in annex 5 the validation of the same is shown and through triangulation, which is a tool used in order to validate the information obtained through different sources, there are five types of triangulation, which include the one of

validation of the data, the researcher, the theory, the methodology and the environment (Script, 2002)

In the following research, the triangulation of the data generated through the interview to the partners of the cooperative present in the study area, and the interviews to the coordinator of the Ciara Foundation, Engineer Arquimedes Santaella, and to the coordinator of the Socialist Agricultural Development Fund (FONDAS), Engineer Yolimar Rodriguez, were carried out. In addition, the agricultural financing programs and plans of NUDE "19 de abril", the technical assistance and the status of execution of the agricultural project developed in the study area were evaluated and examined.

3.8 DESCRIPTION OF THE PROCEDURES FOR THE EXECUTION OF THE RESEARCH.

The description of the procedures was executed in the agricultural production system of the Nucleo de Desarrollo Endogeno 19 de abril, located in the peasant settlement of the same name, in the procedure of execution of the research, aspects and indicators, both qualitative and quantitative, were considered.

The starting point was a diagnosis under a systematic and integral approach of the agricultural production system of the Endogenous Development Nucleus "19 de abril" where the sample was defined, the observation corresponding to the physical and natural aspects of the study area was carried out, as well as the environmental and technical-productive aspects were visualized, as well as other elements that are part of the study unit. The survey and interview were also used, to know elements of social character, since the opinions of the partners of the cooperative and the officials of institutions involved with the agricultural development were considered, so that the procedure of the information included the following stages:

STAGE I

1. Review and analysis of theoretical and legal material such as the Constitution of the Bolivarian Republic of Venezuela 1999, the National Project Simon BoKvar 2007-2013 and some laws, as well as consultations on the web, national and international publications, theses and research papers related to the object of study.

2. Visits to agricultural development institutions such as the National Land Institute, the CIARA Foundation, the Institute for Rural Agricultural Development, the Institute for Integral Agricultural Health, the Fund for Socialist Agricultural Development, for the revision of agricultural charts for the location of the study area and the revision of other materials (constitutive acts of the cooperative, files, technical reports and record cards related to NUDE April 19), to contribute to the justification, problem statement, objectives, population, sample, definition of variables, among others.

3. Elaboration of instruments, with the revision of research texts according to the objectives set.

4. Interviews, review and documentary analysis of community records such as the

community participatory diagnosis, endogenous community development projects and community development plans were carried out, which contributed to the elaboration of the dimensions of the variables and indicators to be evaluated.

STAGE II

1. Meetings and dialogues with the cooperative's members to whom we are interviews were conducted to record information related to social, economic and technical-productive aspects.

2. A characterization of the site was made to obtain information related to the natural physical conditions of the study area including vegetation, soil conditions, water, topography, type of productive techniques and infrastructure present. For this purpose, an observation guide, a geographic positioning system or GPS and the construction of a test pit in the study area were used.

3. Construction of the theoretical reference and research background, as well as the design of the methodological framework.

4. Selection of the information related to the aspects to be evaluated, considering the indicators designed for this purpose.

STAGE III

1. Once the information was processed and the results were obtained, we proceeded to the analysis considering the objectives set according to the system of operationalization of variables by dimension, indicators and sub-indicators. sub-indicators

2. Elaboration of conclusions.

3. Presentation of the proposal with its respective action plan, based on the analysis of results through the design of an integrated agricultural production system as an agroecological strategy in the NUDE "19 de abril".

3.9 DATA ANALYSIS TECHNIQUES.

The analysis techniques applied were bar graphs where various comparisons were established between the different aspects evaluated, also analysis tables were used where the results from the techniques and instruments applied during the development of the research were reflected.

CHAPTER IV

PRESENTATION AND ANALYSIS OF DIAGNOSTIC RESULTS

This chapter presents and analyzes the results obtained in the research. It is a preliminary study that can mean a contribution for those institutions that make decisions for the establishment of agricultural development plans in the country, and in that measure transform the conventional agricultural productive model for a model of systematic and integrated vision, in harmony with the environment and that considers the relationship between man and nature.

In order to respond to the research objectives, we proceeded to analyze each of the dimensions with the indicators shown in Table 3 of the operationalization of the variables. For the physical-natural characterization of the area under study, the environmental dimension was considered, identifying fundamental elements such as:

Environmental dimension

1. Vegetation

In order to identify the type of vegetation characteristic of the agricultural production system of NUDE "19 de abril", a tour of the study area was made, which covers an area of 29 hectares, where a natural vegetation of savannah with shrub and graminiform species was visualized. According to Torres (1996), it is a natural vegetation with little potential for animal production, due to its poor productivity and low nutritional value, since it is deficient in protein, energy and minerals.

The vegetation present in the production system of NUDE 19 de abril presents in some sectors a difference in abundance, observing that in some sectors there has been an anthropic intervention as a result of the agricultural activities that are developed in Table 4 shows the variation of the vegetation in the different sectors of NUDE 19 de abril, where the highest abundance of vegetation is found towards the galena forest zone and the lowest abundance of vegetation towards the agricultural and infrastructure construction zones. Table 5 also shows the main predominant plant species of the production system and the type of vegetation, to identify the plant species we used direct observation, some taxonomical keys and texts of predominant plant species of the non-floodable seasonal savannas.

Table 4 Abundance of vegetation by sectors of the agricultural production system of NUDE April 19.

Sectors	Abundance by plant species
Galena forest	More than 20 plant species
Agricultural and infrastructure construction	Less than 20 plant species

Source: Own elaboration 2009

Table 5 Predominant plant species in the production system of NUDE 19 de abril

Scientific name	Common Name	Type of vegetation
Cassia moschata	Canaffstola	shrub
Xilopia aromatica	Donkey fruit	shrub
Piptadenia peregrina	Yopo	shrub
Copa^fera officinalis	Oil	shrub
Bowdichia virgilioides	Cork oak	shrub
Cassia sp.	Yellow flower	shrub
Trachypogon gallanus	Hairy Straw	Graminiformes
Rinchelintus rosea	Pink Straw	Graminiformes
Panicum maximun	Gamelote	Graminiformes
Trachypogon sp	Stubborn straw	Graminiformes
Sporbolus indicus	Illusion	Graminiformes
Cenchus echinatus	Cadillo bravo	Graminiformes
Andropogon vicornis	Cow tail	Graminiformes
Leucaena leucocephala	Leucaena	Shrubs
Hipparrenia ruffa	Yaragua grass	Graminiformes
Cesamun sp	AjonjoK sabanero	Shrub
Byrsonima crassifolia	Chaparro manteco	Shrub
Petrea arborea	Chaparrillo	Shrub
Caesalpinia coriara	Dividivi	Shrub
Enterolobium cyclocarpum	Expensive	Shrub
Mauritia Flexuosa	Moriche	Palm tree
Cyratella americana	Butter	Shrub
Ciperus ferax	cutter	Ciperaceae
Hyptis suaveolens	Mastranto	Shrub
Tomentose honeydew	Bretonica purple	Shrub

Source: Own elaboration 2009

These results were validated at the inventory level with the work done by participants of the Diploma in Agroecology and Sustainable Development of the Agrarian University of Havana,[3] with the methodology of the Vegetation Inventory Manual of the Humboldt Institute and by other characterizations of vegetation made by officials of the National Land Institute (INTI) in the same production unit of NUDE 19, it was found that they are species typical of savanna ecosystem.

2. Soil

The soil of the production unit has a high proportion of sand in the first horizons and low fertility and low moisture availability, which provides poor plant nutrition. It has a sandy loam texture, with an acid pH (hydrogen potential), low content of organic matter, calcium, phosphorus, sodium, magnesium and potassium. The analysis of the soil characterization of the production system was determined through the review of a soil analysis report carried out by the National Institute of Agricultural Research (INIA-Anzoategui) as shown in Table 6, dated 01-08-2006, where samples were taken at three depths: 0-20 cm, 20-40 cm and 40-60 cm.

Table 6. Soil analysis characteristics.

Hydrogen Potential	Presence of	Presence ofMacro nutrients		Presence of micronutrients		
			P	Ca	Na	Mg
Low (Acid 3.5)	Low (0.42 %)	MuyBajo (0.01 ppm)	Very Low (0.62 ppm)	Low (0.23 Meg/100gr)	T	Low (0.15 Meg/100 g)

Source: INIA Anzoategui 2006.

[3] Fourth level studies at Universidad Bolivariana de Venezuela 2007-2008

A test pit of 60 centimeters deep by one meter wide was also made to determine some physical characteristics of the soil such as: texture, color, soil type, permeability and moisture retention. These soils according to the Law of Land and Agricultural Development (2005), in its Art. 115 are class IV - VI, suitable for livestock development, forestry and some crops, according to Casanova (2003) are considered soils of very low fertility, poor in organic matter and macro and micro nutrients that must be worked with the implementation of agroecological techniques for the maintenance and improvement of conditions.

Table 7 shows the physical characteristics of the soil. physical characteristics

determined through the construction of a test pit.

Table 7 Soil physical characteristics of the agricultural production system of NUDE 19 de abril

Texture	Color	Soil type	Permeability	Drainage
Sandy loam	Light gray	IV-IV	Fast from 1,270 to 2540mm/hour	Rapido

Source: Own elaboration 2010

The soil permeability and soil drainage of the NUDE agricultural production system was determined through field practices and validated with the tables of some soil physical characteristics of the Soil Classification Manual 1988.

3. Hydrography

The agricultural production system of NUDE "19 de abril" has a permanent body of water, which runs approximately two kilometers of linear meters and surrounds the galena forest that forms part of the surface of the study area. This body of water according to dialogue with the partners of the cooperative has decreased its flow and in some parts is dammed and there is also a water storage device (Australian tank), built by the Institute of Rural Development (INDER), in the year 2006, with a capacity of half a million of half a million liters and a deep well of 40 meters deep and with a capacity of 2 liters per second. These data were validated by information provided by Engineer Robert Heredia, CIARA Foundation official, who is in charge of the follow-up and evaluation of the NUDE 19 de abril agricultural development project file.

Table 8 shows the water availability of the agricultural production system of NUDE 19 de abril, which is mainly used for crop irrigation.

Table 8. Hydrography (Water availability)

Natural water body	Rio
Artificial water body	Deep well and Australian tank

Source: instrument applied, 2010

4. Topography

The topography was determined through the relief and slope of the land of the agricultural production system of NUDE 19 April, in this case it was visualized an undulating relief with a medium slope, table 9 shows the values of the degree of inclination of the slope and the relief characteristic of the study. study.

Table 9 Topograffa Elements

Relief	Pending	Degree of inclination
Corrugated	median	3% a 7%

Source: Own elaboration 2010

The slope considered was according to cartographic charts of the National Land Institute of Ciudad Bolivar and the slope was determined with the use of a GPS, the degrees of inclination were taken from the qualification of some physical characteristics of the soil of the soil classification manual 1988.

5. Weather.

According to data obtained from the Aviacion Militar Bolivariana de Ciudad Bolivar Meteorological Station, the average values of temperature, precipitation and evaporation have increased in the last ten years, showing very significant changes. Table 10 shows the average values from 1997 to 2007, recorded by the meteorological station.

Table 10 Climate elements

Elements	Average value year 1999	Average value year 2007
Temperature	27°C	29°C
Evaporation	78.08 mm	80.25 mm
Precipitation	800 mm	700 mm

Source: Bolivarian Military Aviation Meteorological Station Ciudad Bolivar.2010

Regarding the description of the social, economic and technical aspects of production, as stated in the objectives and evidenced in the system of operationalization of variables, it was determined that:

Social Dimension

1. The Social Organization

The social organization of the agricultural production system of NUDE 19 de abril is a grassroots cooperative, formed and linked to national programs, especially with the Vuelvan Caras Mission in 2005, with limited liability, whose associates received agricultural training courses from the Institute of Socialist Educational Cooperation, formerly INCE. Table 11 shows how they are organized in the production system of NUDE April 19. This information was validated by Fernandino Marquez, an official of the Ministry of Popular Power for Agriculture and Land, who was the agricultural instructor for the agricultural frontier program managed by INCE.

Table 11 Forms of social organization

Social organization	**Initial number of members**	**Current number of members**	**Genres**
Cooperative	13	7	4 Women3 Men

Source: Instrument applied 2010

2. Land Tenure

As for the tenure of the land where the agricultural production system of NUDE "19 de abril" is located, they are municipal ejidos, the lots were awarded under the form of gratuitous bailment for a period of 20 years, to the cooperative, as stated in agreement number 132, folios 384 to 386 first protocol, volume VII in ordinary session of the Municipal Council of the Heres Municipality of the State of Bolivar, under number 23 dated October 27th of the year two thousand six, information that was validated with the document of protocolization on behalf of the mayor's office of the Heres Municipality.

3. Social Services

In terms of basic services, NUDE "19 de abril" has transportation, electricity, telephone and good water availability. They have received four agricultural financings from credit institutions such as the Fondo de Desarrollo Agncola Pesquero Forestal y Afines, the Ministerio de Agricultura y Tierra, the Corporation Venezolana Agraria and the Fondo BoHvar. Table 12 shows some elements of the social aspects of the integrated production system of NUDE 19 de abril.

Table 12. Social aspects of NUDE "19 de abril".

Public services	**No. of agricultural financing**	**Forms of participation**	**Financing institutions**
Water	4	Activate	AVC
Phone			FONDAFA
Transportation			MAT
Electricity			Bolivar Fund

Source: Instrument applied 2010

Technical Production Dimension

1. Crop Diversity

Regarding the technical productive aspects, it was found that the agricultural practices developed by the producers in NUDE "19 de abril" are conventional, they do not have crop diversity, they cultivate without any planning, the items cultivated are onion (*Allium schaenoprasum*), cassava (Manihot esculemtum) and beans (Vigna unguiculta), They do not have crop diversity, they cultivate without any planning, the crops grown are onion (Allium schaenoprasum), cassava *(Manihot esculemtum*) and beans (*Vigna unguiculta*), the control of pests and diseases is carried out with the use of chemicals or pesticides, technical assistance and training has not had a follow-up for the implementation on a permanent basis.

2. Soil tillage

Soil tillage is done mechanically, without using any soil conservation practices. According to Casanova (2003), these are soils that have many limitations for agricultural production and must be worked with environmentally conservative technologies and agroecological methods to maintain production, Since they are savanna soils, they are very susceptible to be degraded under a conventional agricultural model where the use of agricultural machinery and other implements can affect the structure and other physical and chemical properties of the soil, Table 13 shows the different aspects that correspond to the technical and productive activities.

Table 13 Technical production elements of NUDE April 19

Crop diversity	pests and diseases - weeds	Soil tillage	Training and technical assistance
Vegetables-cebollin	Insecticides	mechanics	Agricultural producer
Roots-yucca	Fungicides		Cooperativism
Legumes-beans	herbicides		Vermiculture

Source: Own elaboration 2010

Economic Dimension

1. Income Level

Among the economic aspects considered in the study area, the level of income, at the beginning of the project development, was determined to be approximately 60%, since all the members of the cooperative that work in NUDE 19 de abril were receiving the benefit of the corporate advances when there was agricultural production in it.

The income of the NUDE was affected, decreasing by approximately 20%, as shown in Figure 1, which shows the level of income per year.

Graph 1 Income level NUDE April 19- April 19- April 19- April 19- April 19, 2009

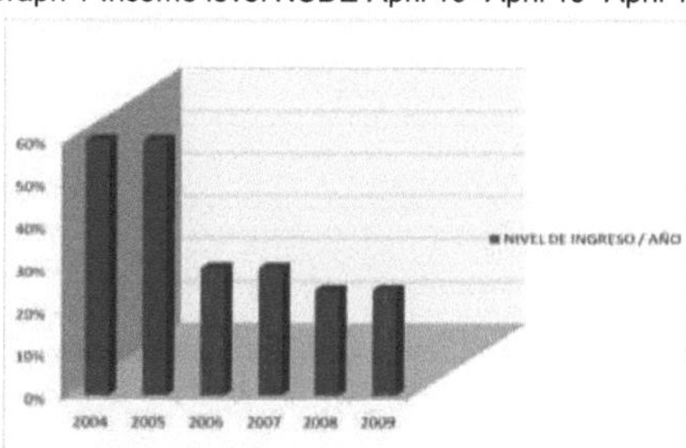

1. Employment

Employment in NUDE "19 de abril" has been decreasing from approximately 70 % in 2004, when the agricultural development project began, to approximately 30 % between

2008 and 2009.graph 2 shows the percentage of employment that has been affected by the project.
established per year.

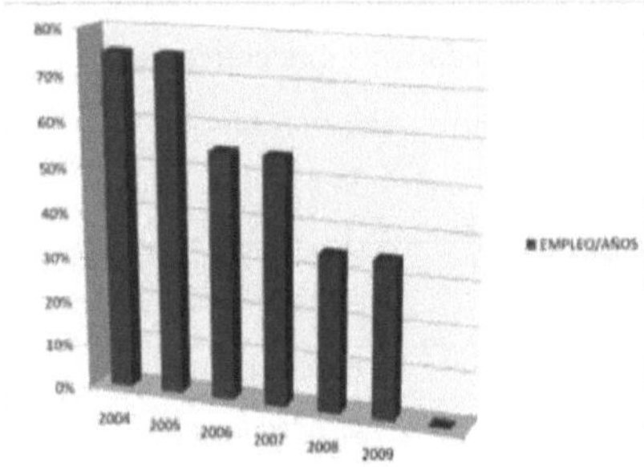

Graph 2. Employment in the NUDE April 19th

2. Sowing area

The planting area from 2004 to 2007 has remained approximately between 5 and 11 hectares. For the years 2008 and 2009 there has been no increase in the area planted for agricultural production, during this period the approximately 11 hectares have been cultivated with permanent species such as fruit trees and short cycle crops such as cassava, beans and some vegetables such as onions. Graph 3 shows the sowing area per hectare per year in the agricultural production system of NUDE 19 de abril.

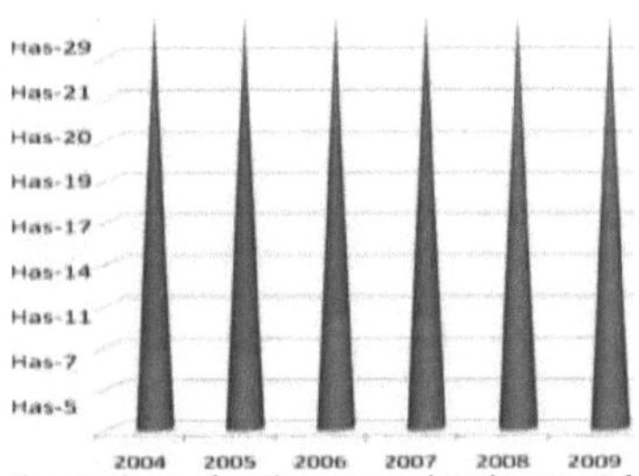

Planting area in the nucleus of endogenous development April 19.

3. Deforestation

The area deforested from 2004 to 2008 has been an area of approximately 21 hectares. It started with 7 hectares deforested in 2004, then increased to 11 hectares in 2006, and by 2008 there are 21 hectares deforested. Graph 4 shows the deforested area per year of the total surface of the production system. production system of theNUDE19of April

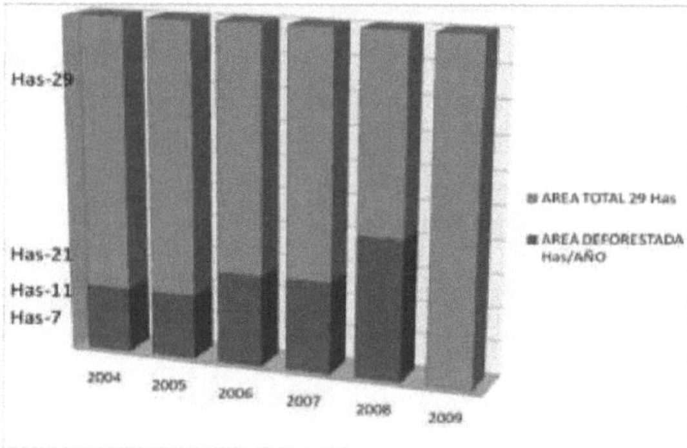

Figure 4 Deforested area of NUDE April 19

Agro-productive dimension

The initiative to promote an agricultural production system under the social figure of an Endogenous Development Nucleus in the peasant settlement "19 de abril" obeys a State policy on agriculture. It was implemented in a centralized and decontextualized manner with the application of a conventional agricultural model that impacted a pristine savanna ecosystem without considering the natural, social, cultural and economic elements. Thus, areas were unjustifiably deforested because they do not have the conditions for agricultural productivity that the endogenous development project assumes, despite the financial measures assumed by the Venezuelan State that generate losses in the economic, environmental and social aspects, and that do not even generate a transformation of the producer's conscience that would result in an improvement in the quality of life, Table 14 shows the socio-productive situation of the production system of theUDE19of April. April.

Table14 Agro-productive situation of the agricultural production system of NUDE 19 de abril

Activities performed	**% of execution**
Deforestation of 21 ha.	100% executed
Deforestation of 3 ha. for paddock construction	Not executed
Deforestation of 1 ha for housing construction	Not executed
Construction of 3 chicken sheds	1 finished shed and 2 sheds in construction
Equipment and supplies for broiler hatcheries	100% delivered
1 Australian dam	Built, not in operation

Acquisition of agricultural machinery	1 Tractor and 1 harrow are used to prepare the land and are rented to other farmers. 1 dusting machine, 1 fertilizer spreader, not used
Financing of 7 hectares of corn	100% financed, part of the harvest was sold to private individuals and part to the industry
Improvement of access roads	100% executed
Deep well drilling	100% realized, in operation, used for the irrigation system.
Construction of 3 km of penetration road	100% executed
3 kilometers of electric service connection	100% executed
Financing of 3 hectares of beans	100% delivered, harvest sold in the local market
Financing of 4 hectares of cassava	100% delivered, harvest sold in the local market
Financing for 1 hectare of milkweed	100% executed, in production

Source: Instrument applied 2010.

CHAPTER V

CONCLUSIONS

The following are the main conclusions of the research, which constitute an overview of the main findings and results of the work, as well as the objectives that were set.

First of all, the negative impact on the natural physical elements (vegetation, soil, water, climate, topography) in the integrated agricultural production system of NUDE 19 de abril, prevailing in this savanna ecosystem, which were not considered for the establishment of this agricultural development project, since these elements define at a specific moment, the best use that can be assigned to an agricultural production unit that allows the establishment of sustainable agroecosystems and the development of crop associations, polycultures or other alternative production systems appropriate to that plant formation, reducing the possibility of the use of toxic products, especially with today's global environmental conditions.

The natural vegetation, typical of the savannah ecosystem where the Nucleo de Desarrollo Endogeno 19 de abril is located, has diminished considerably, as a result of the various conventional agricultural activities carried out in the agricultural production system, favoring the growth of other invasive plant species and the proliferation of some pests, as well as the decrease of the fauna, by not providing them with the necessary conditions to survive. Conditions conditions for to develop.

The soil has been mechanized on several occasions, through the use of agricultural machinery and implements such as harrows and rotary harrows, reducing the topsoil, and therefore the fertility of the soil, depleting it disproportionately, the potential use of the same should be studied and the vocation of these should be considered. The hydrography is given by a permanent body of water whose flow has diminished as a result of the placement of obstacles placed in the normal course, and the construction of water storage devices that are currently inoperative.

In this same order, the social, economic and productive technical aspects of the integrated agricultural production system of NUDE 19 de abril were described, since these elements are fundamental to undertake any type of development project, where it became evident that for a true social transformation process, ideological formation is required to promote true participation and empowerment of the collectives to reach the established goals. The imperious need for a true process of social transformation, which requires ideological formation that promotes true participation and the empowerment of the collectives to achieve the established goals, was evidenced.

Disarticulationand social relevance on the part of members of the cooperative, it is important to achieve a less practical vision of endogenous development that genuinely involves social transformations.

Regarding the economic aspects considered in the variables, it is reported that the level of income and employment of NUDE 19 de abril has declined since the associates stopped receiving the social advance that they received at the beginning of the project, which shows the motivation for economic stimulus, with respect to the planting area and the deforested area, which do not correspond because the planting area has not increased, Despite the deforested area, there is evidence of a lack of awareness and conservationist criteria of the producers who have developed agricultural activities that degrade the environment, as evidenced by the productive techniques they use such as pest control with the use of chemicals, the lack of productive association and diversification, the use of agricultural machinery for tilling the soil, among others.

With respect to the agro-productive conditions of the Nucleo de Desarrollo Endogeno 19 de abril, we found deficiencies, non-compliance and lack of completion of some of NUDE's agro-productive development activities, such as the deforestation of three hectares for the construction of paddocks and the non-use of the Australian tank, This shows that in spite of the state's policies of support to the national agricultural production through its financial entities in NUDE 19 DE APRIL, it has not generated a real agricultural development as it was established.

Finally, we can conclude by affirming the need to eradicate this type of situation and promote agricultural development alternatives that promote a true sustainable agriculture in harmony with the environment, where the relationship between man and nature is harmonized.

CHAPTER VI

FORMULATION OF THE PROPOSAL

6.1 PRESENTATION

Nowadays, the land-soil resource plays a preponderant role not only in the quality of production, but also in the preservation of natural resources, especially water and vegetation, which must be taken into account in order to undertake any type of agricultural agricultural development, particularly if the agricultural activity is exploited under a high-input or conventional technical approach, as it has been done so far in our country, which does not guarantee the expected sustainability of agricultural development, that is why the only possible development that remains is the one based on the frank cooperation between man and production relations that humanize society.

In order to collectively aspire to food security for our country, based on a sustainable development, enhancing the agricultural activity, this research proposes to establish the Integrated Agricultural Production Systems as an agroecological alternative in savanna ecosystems, where the importance of natural resources and the many particularities that make up our agroecosystems are recognized.

Integrated agricultural production systems correspond to an agroecological model that breaks with conventional predatory practices and that have generated problems related to land degradation and its excessive use, compaction, irrational application of agrochemicals, leading to an alteration of the natural ecological balance by reducing or disappearing vegetation factors, soil fertility, biodiversity, nutrient recycling and the reduction of microbial activity.

6.2 OBJECTIVES OF THE PROPOSAL

6.2.1 General Objective:

To propose the establishment of an Integrated Agricultural Production System as an agroecological alternative in the Endogenous Development Nucleus "19 de abril" Heres municipality of the State of BoHvar.

6.2.2 Specific objectives

Develop activities to sensitize the members of the cooperative for the activation and construction of construction collective oriented to social organization and integral agricultural development.

Execute activities Theoretical and practical oriented to the use of agroecological techniques in the production unit of NUDE April 19 to transform the practices of the practices practices conventional.

Integrate the different subsystems that make up the production system of Nude 19 de abril in order to establish the Integrated Production System. Production System Production System.

6.3 JUSTIFICATION

Traditional agricultural systems are generally characterized by a low level of technological development, extensive land use and, consequently, low productivity of land and labor, in particular shifting cultivation and savanna agriculture. Nowadays, population growth and competition from cash crops make it necessary to increase production, which leads to land depletion and, in the long run, to a lack of food and the desertification of the savannah.

The Integrated Agricultural Production Systems seek to improve the living conditions of farming families and reduce environmental depredation, because it allows the combination of a wide range of activities, which a human group organizes, directs and carries out according to its objectives, culture and resources, using conservationist practices with responses to the environment.

Figure 2 shows an outline of the Integrated Agricultural Production Systems and the components that justify and visualize them as functional structures in the construction of a new agricultural paradigm, articulating environmental, economic and social processes that determine the forms of appropriation, construction and transformation of new styles of sustainable development. sustainable and agroecological development styles.

Figure 2 Rationale for Integrated Agricultural Production Systems.

6.4 PROPOSAL RATIONALE

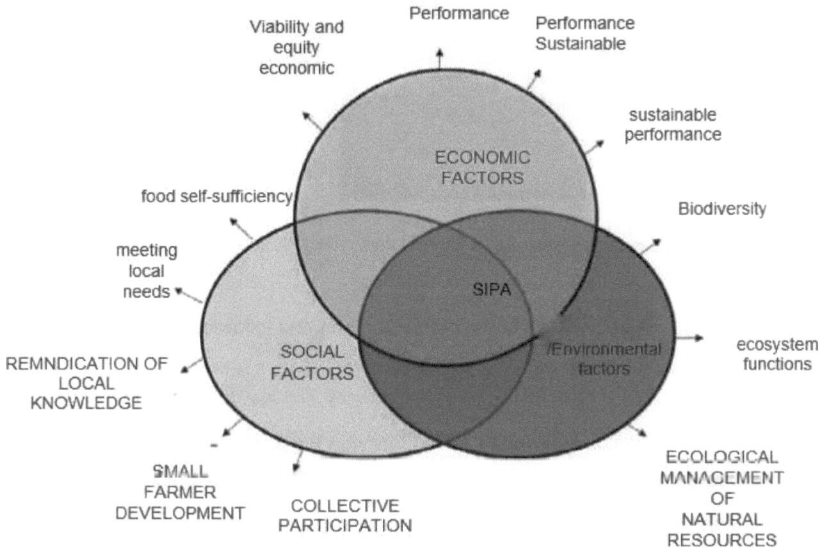

The proposal is based on coherence, integralityand complementarity that underlies the Integrated Agricultural Production Systems, including aspects between technical, productive, environmental environmentaland

The following figure 3 shows schematically how the integrated systems of agricultural production are integrated and supported by these three elements, which are the basis of ecological sustainability, social equity and cultural diversity, reaffirming the rights of the communities to self-management of their natural resources, revaluing the ecological productivity and traditional practices of the resources.

which characterize the SIPAs as an alternative model of technology transfer and agroecological extension.

Figure 3: Integrated Farming Production Systems Foundation

Execute theoretical-practical activities oriented to the use of techniques agro-ecological practices in the production unit of the NUDE19 of April for transform them agricultural practices conventional.	Realization of theoretical and practices in the aspects partner technician with the partners of the cooperative in the agricultural production of NUDE April 19.	-To carry out training workshops, oriented towards collective work, linked with the social organization asthe cooperatives, cooperative companies social production , communal council and units of agroecological production with the accompaniment for the conformation of the same. -The projection of videos from experiencesfrom conformationde social organization and agroecological plots. -Development of technical-productive activities, such as de monstrations of methodsand experiences gained in at other units agroecological production.	Six months	Facilitators from institutions with responsibility in agricultural development, partners of the kamaropa cooperative	Technical. Productive Human. Didactic. Physical
Integrate them different subsystems that make up the the system productionof Nude April 19 to establish the System Integrated from Agricultural Production.	Constitute them different subs ystemsthat will be integrated into integrated production production agncola.	-Develop them different productive subsystems on study unit. -Application of the different agro-ecological practices that help to improve the maintenance and integration of each subsystem.	1 year	Technicians with knowledge in the area agroecologica, partners of the Kamaropa cooperative, governmental and state institutions of support agricultural sector.	Technical. Productive Human. Physical. Didactics

Table 15 cont ....

Source: Own elaboration 2010

6.7 METHODOLOGY FOR THE DEVELOPMENT OF THE PROPOSAL

For the development of the proposal, the system approach is proposed, to facilitate the study and understanding of the agricultural activities of the NUDE "19 de abril" through the implementation of participation tools that will allow to know and recognize the integrality or interrelations between the parts of the NUDE "19 de abril", as well as the state of protection and production of the resources, the organizational levels of incidence in the community, the integrality or interrelations between the parts of the NUDE "19 de abril", as well as the state of protection and production of the resources, the organizational levels of incidence in the community, for this reason it is inserted in the methodological dynamics of the Participative Action Research and dialectics.

This methodology is established as a procedure that seeks transformation and collective learning, which considers the nature of individuals and their relationships with the physical and social world, according to Salazar 1999 and Fals 2000, it is a methodology that gives rise to emancipatory processes, which seeks to achieve a better possibility of transformation of a given situation and allows the development of creative thinking through learning by doing, with the use of techniques of participation and collective actions.

6.8 DETERMINATION OF RESOURCES

6.8.1 HUMAN TALENT

For the development of the Integrated Agricultural Production System, we count on the cooperative's associates, support institutions for social organization, productive technical training, education and others (CIARA, INSAI, INDER, UBV, FUNDACOMUNAL).

6.8.2 MATERIAL RESOURCES

The materials to be used are divided into items of a technical-productive nature and those of an educational nature.

Technical and productive character:

1. Seeds of the different crops to be developed in the vegetable production subsystem.
2. Five cows and one bull for the development of the bovine animal part.
3. Five sows and a stallion.
4. 1OO fry for the start of the fish hatchery.
5. Nursery bags.
6. Beekeeping boxes.

Educational character:

1. Ream of paper
2. Acoustic blackboard.
3. Scoreboards.
4. Video ben.
5. Drafts.
6. Paper sheets.
7. Television and DVD.

6.8.3 FINANCIAL RESOURCES

In order to obtain financial resources for the proposal, it is being contemplated to refinance the loan that NUDE "19 de abril" has through the new policy for the re-promotion of the Nucleos de

Endogenous Development at the national level by adapting the initial project towards the establishment of Integrated Agricultural Production Systems with an agroecological approach to promote food security and develop sustainable agriculture with preservation of the ecosystem and natural resources.

6.9 BUDGET

The estimated budget to develop the proposal will be represented by a refinancing of 50% of the amount of the initial NUDE "April 19" loan.

6.10 EXECUTION SCHEDULE

The execution schedule of the proposal is contemplated to be carried out over a period of two years with the following activities.

1. PART:

   1st year.

   a) Developmentof plant

subsystem which comprises the establishment of crops such as vegetables, pastures, leguminous plants, fruit trees, forestry, and tubers, as well as source food human and animal .

b) Development of the animal subsystem starting with the production of swine and previous recovery of the pig farms, and the cattle production.

c) Development of fertilizer production subsystems organic waste through vermiculture and the recycling of animal and vegetable waste.

d) Development of beekeeping.

2 PART

2nd year.

a) Development of the fish farming subsystem.
b) Development of the Poultry subsystem.
c) Development of the value-added subsystem

6.11 FEASIBILITY OR VIABILITY OF THE PROPOSAL

The proposal to establish an Integrated Agricultural Production System with agroecological bases is feasible and viable since it is about putting into practice a new model of agricultural production, conservationist of the environment and natural resources as well as taking advantage of the potential of the sector to promote endogenous development and conclude the development of a socio-community project, In addition to guaranteeing food sovereignty and generating in agricultural producers the concept of the need to conserve natural resources through the implementation of agroecological practices that help minimize the use of agrochemicals, minimize soil degradation, and preserve biodiversity.

This approach considers ancestral knowledge, making a rational use of agroecosystem resources, protecting and conserving water bodies, diversifying production, maximizing the use of internal resources of the production unit and minimizing the use of external resources, and last but not least, considering the human being as the fundamental element to generate any change.

6.12 BENEFITS OF THE PROPOSAL

1. Actions taken with full social participation by the communities themselves generate profound changes in the rural context, since they are focused on the development of participatory projects of small and medium agricultural production that help the communities to achieve food autonomy.

2. In promoting the creation of Integrated Agricultural Production Systems, the technical aspects of production, social organization, cultural values, environmental preservation and biodiversity must be harmonized in order to biodiversity for the use of the local potentialities by promoting the conversion of a system of local conventional monoculture or diversified diversified system or agroecological, in order to achieve quality production and its food self-sufficiency.

3. The agroecological approach oriented towards Integrated Agricultural Production Systems assumes that what is essential is that the person, family, community or organization adapts to their own reality, the strategies to meet their needs and enhance their own virtues and wealth, In other words, it is a question of seeing the establishment of FAPS as a socio-productive tool, which favors balance and complementarity between the protection and production of resources, production and organization, organization and advocacy, advocacy and empowerment of individuals and communities.

BIBLIOGRAPHICAL REFERENCES

Altieri, M. (1999) Bases cientificas para una agricultura sustentable Montevideo. Uruguay.

Altieri, M. (2000) "iPor que estudiar la agroecolog^a?", in Agroecolog^a y Agricultura Sostenible. Modulo I. Agroecologia: Bases Historicas y Teoricas. Centro de Estudios de Agricultura Sostenible del Instituto Superior de Ciencias Agropecuarias de la Habana.

Altieri, M. (1997) Fundamentos filosoficos y cientfficos de la agroecolog^a. Santiago de Chile. Chile.

Altieri, M. and Letourneau S. (1994) Agroecological perspectives. Biodiversity and agroecological design. Integrated pest management and agriculture. Costa Rica.

Arias, F. (2006) The Research Project. Introduction to scientific methodology. Fifth edition. Editorial Episteme. Caracas Venezuela.

Badillo, J. (1995) Taxonomic keys of plant and animal species.

Balestrini, M. (2006) Como se elabor se elabora el proyecto de investigacion. Caracas.

Casado, G. (1997) Towards the development of a participatory methodology for agroecological transition at the local community level. Cordova Spain.

Casado, G. (1999) La investigacion participativa en agroecolog^a; una herramienta para el desarrollo sustentable. Consorcio Centro de investigacion y formacion de agricultura ecologica y desarrollo rural. Santa Fe. Granada.

Casado, G. (2004) Introduction to agroecology as sustainable rural development. Mundi prensa Madrid.

Casanova, E. (2003) Sustainable agriculture and soil fertility.

Centeno, J. (2000) Forests of Venezuela. El desarrollo Forestal de Venezuela. Merida.

Chauran N, Gomez C and Nieves A. (2001). "Characterization of the agricultural production systems in the locality of La Guanota, Caripe Municipality, Monagas State". Memorias VII Congreso Cientifico UDO, Nucleo Nueva Esparta.

Comerma, J. and Paredes, R. (1998) Main limitations and agricultural potential of the land in Venezuela. Tropical Agronomy.

Constitution of the Bolivarian Republic of Venezuela (1999) Official Gazette of the Republic, 36.860. December 12, 1999.

Fals, B. Participatory action research. Accion y Conocimiento. Colombia. 2000.

Fernando C. (2008).agricultural, livestock, forestry, aquaculture and fishery production systems. Research work. Lmeas de investigación de sistemas y producción. Mexico.

Funtowic, S. (1994) Epistemologia poHtica: Ciencia con la gente. Buenos Aires: Centro editor de America Latina.

Garda, R. (2004) Al andar se hace camino. Guia metodologica para desencadenar procesos autogestionarios alrededor de experiencias agroecologicas. Colombia.

Garrido J (2007). The research project. HoKstic understanding of methodology and research. Ediciones Quiron. Caracas Venezuela.

Gliessman, S. (2002) Agroecolog^a: Procesos Ecologicos en Agricultura Sostenible. Turrialba, Costa Rica.

GuionL . (2002). Triangulation: Establishing the validity of qualitative studies. Accessed August 5, 2010. Available at http// edis.ifus.ufl.edu.

Guzman, M (2000) Las metodologias participativas de investigación: Un aporte al desarrollo local endogeno. Pamplona.

Hurtado, J. (2008) The Research Project. Methodology of the hoHstical research. Fourth edition. Bogota.

Kaimowitz, D. (1997) Tecnolog^a Agncola Sostenible: retos institucionales y metodologicos dos estudios de casos en Colombia. Serie documentos de

discussion on Sustainable Agriculture and Natural Resources. IICA. San Jose. Costa Rica.

La Roche A. (2006). Characterization of livelihood strategies and agrosilvopastoral technology of households in Esparza Costa Rica. Degree thesis. Education and conservation program.

Land and Agrarian Development Law. 2005. Extraordinary Official Gazette of the Bolivarian Republic of Venezuela No. 5890, July 2008.

Organic Law on Agrifood Security and Sovereignty. 2008. Official Gazette of the Bolivarian Republic of Venezuela N° 5.891, July 2008.

Leyva, A. and Pohlan, J (2000) Agroecology in the tropics. Examples from Cuba. Plant biodlversity how to conserve and multiply it.

Manual of soil classification. 1988 Shell Foundation. Venezuela.

Nunez, M (2007) Por una poHtica agroecologica .La agroecolog^a en la Soberania Agroalimentaria Venezolana. Merida.

Nunez, M (2002) Proposal for sustainable rural development. Venezuela. Caracas.

Nunez, M (1999) Alternativas tecnologicas de producción agroecologica. Manuals of environmental education and training. D.F. Mexico.

Food and Agriculture Organization of the United Nations (1998) Evaluation of land for irrigated agriculture. Soil Bulletin N55. Rome Italy.

Simon National Project First Socialist Plan. Development Economico Social de la Nacion. Caracas, September 2007.

Ram^ez T. (2007)How to to make a proyectode Caracas. Editorial Panapo.

Ram^ez, T (2007) How to make a project research. Editorial Panapo. Caracas.

Restrepo, J (2000) Agroecologia. Universidad Nacional de Colombia and Fundacion para la Investigacion y el Desarrollo Agncola. Center for Agricultural and Forestry Development, INC. Colombia.

Revolutionizing the Venezuelan Countryside. Towards sovereignty agri-food. CIARA Foundation 2008.

Rodriguez, R. (2007) Agricultura Tropical con Enfoque Humano y Vision Sistemica. Villavicencio Colombia.

Salazar, M. Participatory Action Research. Beginnings and Development. Adult Education Library. Tachira. Venezuela. 1999

Sarmiento E. (2007). Characterization of productive systems in the rural area of Guavio. Research work of the Natural Patrimonial Agreement.

Corpoguavio Emgesa. Conservation and Agribusiness Production in the area of influence of the Guavio. Colombia

Sevilla, E. (2001) Agroecology and Sustainable Rural Development: A Proposal from Latin America.

Suquilanda, M. (1999) Agricultura Organica: Altemativa tecnologica para el futuro. Foundation for Rural Development. Quito Ecuador.

Torres G. (1996). Management of Venezuelan savannas, problems and perspectives. Serie D. San Fernando de Apure. Venezuela.

Veliz A. (2009) Como hacer y defender una tesis. Caracas. Editorial Texto.

Vilcher, A. Tools for Entrepreneurs. 2008.

More Books!

yes
I want morebooks!

Buy your books fast and straightforward online - at one of world's fastest growing online book stores! Environmentally sound due to Print-on-Demand technologies.

Buy your books online at
www.morebooks.shop

Kaufen Sie Ihre Bücher schnell und unkompliziert online – auf einer der am schnellsten wachsenden Buchhandelsplattformen weltweit! Dank Print-On-Demand umwelt- und ressourcenschonend produzi ert.

Bücher schneller online kaufen
www.morebooks.shop

info@omniscriptum.com
www.omniscriptum.com

OMNIScriptum

MIX
Papier aus verantwortungsvollen Quellen
Paper from responsible sources
FSC® C105338

Printed by Books on Demand GmbH, Norderstedt / Germany